SKILLED AND VOCATIONAL TRADES

BECOME A WELDER

by Carla Mooney

BrightPoint Press
San Diego, CA

BrightPoint Press

an imprint of ReferencePoint Press, Inc.
Printed in the United States

For more information, contact:
BrightPoint Press
PO Box 27779
San Diego, CA 92198
www.BrightPointPress.com

LIBRARY OF CONGRESS CATALOGING-IN-PUBLICATION DATA

Name: Mooney, Carla, author.
Title: Become a welder / by Carla Mooney.
Description: San Diego, CA: ReferencePoint Press, 2027 | Series: Skilled and vocational trades | Includes bibliographical references and index. | Audience: Grades 7–9
Identifiers: ISBN 9781678214661 (hardcover) | ISBN 9781678214678 (eBook)
The complete Library of Congress record is available at www.loc.gov.

CONTENTS

AT A GLANCE

- Welding is a process that joins two or more metal pieces. The metal edges are melted together to create a strong bond.

- Welders help make many things. Their work is used in airplanes, rockets, bridges, buildings, and more.

- Welding trade schools and technical programs give students hands-on practice. Students spend many hours in workshops using real welding equipment.

- Welders can learn on the job through apprenticeships. They get paid while working with experienced welders.

- Important skills for welders include attention to detail, problem-solving skills, and physical strength and stamina.

- Welders work across many industries for manufacturers, trade contractors, and repair shops. Some are self-employed.

- Jobs for welders are expected to grow by about 2 percent from 2024 to 2034. But welders will also be needed to replace those retiring or switching careers.

- Automation is changing the way welding is done. It may reduce the need for some welders, but welders will still be needed for complex projects.

WELDING FOR OUTER SPACE

Kody Chaney is a welder. A welder uses heat and other tools to join pieces of metal together. He has been working on welding projects since he was 5 years old.

Chaney's father worked as a welder for more than 40 years. Growing up, Chaney helped his dad with many projects. They repaired farm equipment. They fixed cars and boats for family and friends.

Welders work in a variety of industries, building infrastructure, cars, tools, and more.

Chaney knew from a young age that he wanted to be a welder just like his dad.

Now Chaney works as a welder for Northrop Grumman. Northrop Grumman is a company that makes technology. Workers make airplanes that fly without pilots. They build rockets. They make systems for the military.

Chaney is the lead welder for part of the Gateway Space Station. This is a space station that will be used by the National Aeronautics and Space Administration (NASA). Chaney is building the Habitation and Logistics Outpost (HALO). This is where astronauts will live while at the space station.

Chaney and his team are welding parts of the HALO's thermal control system.

This system will keep the astronauts at a safe temperature on the space station. The work is challenging. Chaney and his team are welding more than 1,000 weld joints on the HALO. A weld joint forms when two or more pieces of metal are welded together. Chaney's team must make sure the welds are strong. The joints need to hold together

The Gateway Space Station will orbit the Moon.

in outer space. The astronauts' lives depend on it.

Chaney regularly visits his old high school. He talks to students about working in the welding **industry**. He says:

> *I felt like I was fortunate enough to be given opportunities and chances to explore the* ***aerospace*** *industry and use my skills in a really cool way. I want kids to feel inspired to do the same and dream big.*[1]

BUILDING THE WORLD

Welders help build many everyday items. They make cars. They build bridges. They make complex machines out of metal. They use heat and welding tools to join the metal parts together.

Some welders build new structures. Others fix existing structures.

Welding can be dangerous. Welders must work with very hot materials. They must wear safety gear to protect themselves. This gear includes helmets and gloves.

Welding takes focus. It also takes practice. But the job is worth it. Welders are needed around the world. Their skills help build the strong structures and machines that people depend on.

CHAPTER ONE

WHAT DOES A WELDER DO?

The world is full of metal objects. Many of these objects are built by welders. Welders are skilled workers. They make sure the objects people use are strong and safe.

Welding is a process that joins two or more metal pieces. To do this, welding uses heat and pressure. The edges of the metal pieces melt. The melted edges combine. They form a strong bond.

Welders are in high demand, meaning workers can often find jobs quickly.

Welders work with several types of plastic, including PVC, acrylics, and nylons.

Welders work with a wide range of metals. One common metal is steel. This is a mix of iron and carbon. Welders also work with aluminum. This metal is strong. But it is also light. Welders may also work with plastic.

STARTING THE PROJECT

Welders start projects by looking at blueprints. These are drawings. They show the welders what they need to build. Blueprints show the different parts welders must make. They show the size of each part. They show the shape and position of each piece.

Welders must follow blueprints carefully. Otherwise the project might not work. The final product could be dangerous.

Welders then prepare their supplies. This includes their tools. It also includes the metal they will be welding. They clean the metal. They remove any rust or paint.

Then welders make the pieces the right shapes. They may need to cut the pieces. Welders can do this using grinders or cutters. These are machines that can cut through metal. Once everything is prepared, the welding can begin.

WELDING

Metal melts at high temperatures. So welders must get their metal very hot. This requires special tools.

One important tool is an arc welding machine. This machine melts metal. It uses electricity to create an electrical arc. This looks like a bright spark. This spark is very hot.

The welder aims the spark at the edges of metal pieces. The heat melts the metal. The welder moves the spark along the metal edges. The parts that are no longer touching the spark cool. They become solid again. This forms a strong bond. The bond holds the pieces together.

Welders check their work when they are done. They make sure the bonds are safe

Arc welding machines heat metal to about 6,500°F (3,600°C).

and strong. They check for cracks. They make sure there are no holes. Sometimes they use tools to test the pieces. They may use sound waves or X-rays to make sure the products are solid. This keeps everyone who uses the products safe.

Welders must also keep themselves safe. They must follow safety rules. Welders have

WHAT PERSONAL PROTECTIVE EQUIPMENT DO WELDERS NEED?

- HELMET
- GOGGLES
- RESPIRATOR
- FLAME-RESISTANT CLOTHING
- GLOVES
- BOOTS
- EAR PROTECTION

Personal protective equipment keeps welders safe from burns, electrical shocks, and fumes.

dangerous jobs. They work with flames. Their tools can burn their skin. Bright sparks can hurt their eyes. So welders must wear protective equipment while they work. Helmets and gloves protect their skin. Safety glasses protect their eyes. Welders must also make sure the air stays clean. Otherwise dangerous **fumes** could hurt their lungs.

MANY TYPES OF PROJECTS

Welders work on a wide range of projects. They help build bridges. They work on ships. Some welders repair broken objects. They may fix damaged cars.

Some welders work on pipelines. These are long pipes that carry liquids. Some carry water. Others carry gas or oil.

Pipelines need to be strong. Otherwise they could leak. Welders make sure that does not happen.

Some welders work at big construction sites. These sites are for massive projects. This is where bridges and buildings are made.

Construction sites can be challenging workplaces. The sites are usually outdoors. So welders must deal with all kinds of weather. They may work in high heat or freezing cold. Welders also may have to access uncomfortable places. This includes very high or tight areas.

Other welders work inside factories. These places make smaller projects. Welders may make machines. These machines include cars.

When cars are damaged in crashes, welders can remove broken metal pieces and replace them with new parts.

Welders can also work in repair shops. These places fix machinery. Jo Mackiewicz works in a repair shop. She is part of a team of welders. She works on a variety of projects. Mackiewicz says:

*I've **fabricated** skis for the machines that make the rumble strips in the road, shepherd's hooks for bird feeders, fence poles, and stainless-steel lampshade frames. I've repaired trailers, wheelchair ramps, office chairs, and lawn mowers.*[2]

Welding is a challenging career. It requires attention to detail. It also

Underwater Welders

Underwater welders fix and build metal structures underwater. They work on ships and underwater pipelines. Some underwater welders repair broken parts. Others install new parts. Still others cut metal pieces underwater. These welders must learn both welding and diving skills. Their jobs can be even more dangerous than traditional welding jobs.

requires patience. Working with high heat and electricity can be dangerous. But the job is also rewarding. Welders build important machines. They can create entire buildings. They see the results of their hard work every day.

CHAPTER TWO

WHAT TRAINING DO WELDERS NEED?

Welding is a tricky job. So welders need training. They must learn how to do their jobs right. This keeps the people who use their products safe. It also keeps the welders themselves safe.

Most employers require welders to have high school diplomas or equivalent degrees. High school teaches students important skills. These skills include reading and writing. They also include math

People can learn to become welders in less than a year.

Some high schools partner with trade unions, allowing students to mark their time in the classroom as work experience.

and science. These skills help welders in their careers.

Reading and writing skills help welders understand instructions and safety rules. Welders use math skills to follow blueprints. They must know how to correctly measure materials. They must be able to calculate angles. Science classes teach students about the properties of metals. They also teach students about how heat and electricity affect metals.

Some high schools offer shop classes. These classes teach students about trade jobs. These are skilled, hands-on jobs. Students may learn to read blueprints. They might even practice making their own blueprints.

Some high schools have specific classes in metalworking. These classes give students practice cutting and shaping metal. Students learn to join metal pieces. Some of these classes even teach basic welding skills. Students learn about different welding processes and techniques. They practice using real welding equipment.

Jason Hill teaches welding at a college. He took his first welding class in high school. He took it hoping it would be an easy class. But it led him to his career. Hill says, "Welding came easily to me, so I took to it and ran with it. It was like breathing. I had to do it all the time."[3] Now he gives other students that opportunity.

TRADE SCHOOL AND COMMUNITY COLLEGE

After high school, many future welders go to trade schools. These schools offer programs in welding. The programs usually take between 6 months and 2 years to complete. Future welders may also go to community colleges. These schools offer 2-year associate degrees.

Trade schools prepare students for careers in construction, plumbing, welding, and more.

Students in trade schools and community colleges go through technical training. They practice welding. They learn different welding techniques. They practice using different welding tools. These tools include welding machines. They also include torches and clamps. Students learn how to read blueprints. They practice geometry and measurement skills.

They also learn how to protect themselves with safety equipment.

Welding training programs give students opportunities for hands-on practice. Students spend many hours in workshops using real welding equipment. Hill explains that hands-on practice is essential to any welding training program. He says:

Most welding apprenticeship programs require apprentices to be at least 18 years old.

We give students the experience and the knowledge so that they can understand . . . the different types of welding, how they can use those various techniques, and develop their skills to find their place in the workplace.[4]

LEARNING ON THE JOB

Another way to become a welder is through an apprenticeship. An apprenticeship pairs beginners with experts. Apprentice welders usually work full-time. They often take classes as well. Most get paid for their work. They earn money while they learn.

Apprenticeship programs typically last 3 to 5 years. **Unions** and trade schools often offer apprenticeship programs. So do some employers. Most programs require each apprentice to have a high school diploma or equivalent. Some programs prefer apprentices to have some basic welding knowledge and skills. Others require each applicant to pass a basic welding test. Those who complete their apprenticeships have real-world experience. They are ready to work independently.

EARNING LICENSES AND CERTIFICATIONS

Many welders get certifications. A certification shows that a welder has strong skills. It proves that a welder can perform a certain type of welding. Most get certifications from the American Welding Society (AWS). Having an AWS certification

The AWS offers certifications in teaching, inspecting, and supervising welding.

can help a welder get a better job. Certifications also help welders earn more money. Some employers require employees to have specific certifications.

Many welders choose to earn certifications under the AWS Certified Welder Program. The program includes a test. The test usually involves welding a piece of metal in a specific way. Then the piece is checked to ensure it is strong and correctly welded. Those who pass receive certificates for that technique.

Some states require welders to be licensed. Licensing requirements vary by location. Welders should contact their state or local governments for more information.

Welders may also be required to complete other training. The Occupational

Welders in Motorsports

Some welders work for racing teams. Motorsports teams in the National Association for Stock Car Auto Racing (NASCAR) and IndyCar hire welders to build and repair their cars. Welders build frames and customize cars. They quickly repair broken parts. During races, welders work fast to get damaged cars back on the racetrack.

Safety and Health Administration (OSHA) is a government agency. It works to keep employees safe. OSHA requires welders to complete safety training. This lowers the risk of welders being hurt while working.

ONGOING TRAINING

Welding is always changing. New tools and welding methods are being created. These innovations make work faster and safer. The changes also mean welders must always keep learning.

Welders may join organizations to learn about new techniques. Some take extra classes to learn new tips. Others earn certifications. Some attend safety training classes to learn about new rules. This keeps welders up to date in their field.

CHAPTER THREE

WHAT IS LIFE LIKE AS A WELDER?

Working as a welder is challenging. The job is physically demanding. It is also dangerous. Welders work in extreme conditions. But many welders think the job is worth it.

Most welders work full-time. This is about 40 hours per week. But welders may often work overtime to finish projects. Sometimes a job must be finished quickly. So welders

Some welders travel around the country for work.

may have to work in the evenings or on weekends.

Welders in manufacturing plants often work in shifts. These plants often operate all day. Different welders come in at different times. Someone is always working.

Some plants have two or three shifts each day. A shift can last 8 to 12 hours. One welder might start early in the morning. Another might work overnight. Welders must be flexible. Their schedules can change depending on the project.

Janet Davis is a welder. She usually works 4 days a week. She gets to work at 4:00 a.m. Then she works until 2:00 p.m. But sometimes she has to work more to finish a project. She says, "There have been times where we've worked 5, 6, or

Welders who work overnight often get paid more than those who work during the day.

7 days a week."[5] She explains that her workload changes depending on what customers need.

Davis describes a typical day at her job. "I clock in at 4:00 a.m. and get assigned my job for the day. We have a computer system, so I log into the computer and prepare for my job. Then I get set up; I get my [safety equipment] and tools out, and then I start working. I am welding the entire time I am there. . . . At the end of the day, I clean up and log out, and I'm out for the day," she says.[6] It is a long day, but

Welding outdoors can be dangerous. Rain can lead to electric shocks. Wind can cause sparks to start fires.

welders take pride in their work. They help build important objects that people use every day.

WORK ENVIRONMENT

Welders must be able to work in many conditions. Some work indoors. They might work in hot factories or workshops. Other welders work outdoors. They may have to work in the hot sun. Sometimes they work in heavy rain. They might even work in snow.

Workers must also be able to work in awkward positions. Sometimes welders work high up on **scaffolds** or platforms. They may have to squeeze into tight spaces to reach the metal they need to weld. They might need to kneel or stretch to reach the area they must weld. This makes the job physically demanding.

WHO EMPLOYS WELDERS?

The Bureau of Labor Statistics (BLS) is a government agency. It tracks information about US jobs. The BLS has information about welders. It reported that there were about 457,300 welding jobs in 2024.

The BLS also tracks where people work. It said that many welders worked for manufacturers in 2024. They built machines such as cars. Other welders worked for trade contractors. These companies build big structures. This includes bridges and buildings. It also includes pipelines and roads.

Some welders are self-employed. This means they work for themselves. They take jobs from many different customers. This is called freelancing.

SKILLS WELDERS NEED

Welders need many skills to do their jobs well. One important skill is being detail-oriented. Welders must pay close

Welders on Set

Welders work behind the scenes on movies and television shows. The job can be fast-paced and exciting. Welders work with a crew to build sets. They also make props. They must be good at making many things. Everything from prop weapons to fancy cars for chase scenes requires welders.

Some welders work hundreds of feet above the ground.

attention to their work. Each weld has to be strong. A small mistake can make a structure unsafe.

Welders also need physical strength. They often have to lift heavy pieces of metal or equipment. They must have **stamina**. They have to work for long hours.

Good hand-eye coordination is also an important skill. This helps welders make **precise** welds.

Spatial orientation skills are helpful for welders. These skills allow welders to read blueprints and understand how the pieces fit together. They should be able to imagine what the finished structures should look like.

Welders should also have problem-solving skills. Sometimes problems occur on the job. Welders must come up with creative solutions to solve the problems.

CHAPTER FOUR

WHAT IS THE FUTURE FOR WELDERS?

Welders are needed around the world. So welding is a good career choice. Welders will always be needed to make new machines and structures. They will also be needed to fix old things.

The BLS predicts how job demand will change over time. The agency expects jobs for welders to grow by about 2 percent from 2024 to 2034. This growth is slower than the average for all jobs. But it still means

In 2024, the median pay for a welder was $51,000 per year.

Some welders become inspectors. They make sure welds are done correctly.

many welding jobs will be available. The BLS predicts there will be 467,200 welding jobs by 2034.

The BLS estimates that there will be thousands of job openings for welders each year. It predicts that there will be about 45,600 open welder jobs each year. Most of these job openings are due to older welders retiring. Some may also change jobs. New welders will be needed to fill these positions.

WHERE WELDERS WILL BE NEEDED

Welders will be needed in several areas in the coming years. One important area is repairing aging **infrastructure**. This includes bridges and buildings. It also

Women in Welding

Most welders are men. The BLS reported that only 6 percent of welders in 2024 were women. However, more women are choosing welding as a career. Organizations such as Women Who Weld and the AWS have programs to encourage women in welding. These organizations offer classes and scholarships. Some even offer mentors.

includes tunnels and water pipes. Many of these structures are old. They will need repairs to stay safe. Welders will be needed to fix metal parts on these structures. Welders will need to build new supports. Their work will keep these structures safe and strong.

Welders will also be needed in manufacturing plants. Workers in factories make many metal products. They make machines. They also make tools. Welders help connect and shape the metal pieces needed for many products.

THE IMPACT OF AUTOMATION

Automation may affect welding jobs in the future. Automation is the use of machines to do jobs. These machines include robots.

It is cheaper for companies to use robotic welders than human welders.

Robots already help with welding tasks in some places. They can weld the same part repeatedly. Robotic welding can be faster than human welding. It can also produce fewer errors.

Automation may reduce the need for some welders. But human welders will still be needed. Human welders program robots. They monitor automated systems.

Welders in the future will likely need to learn to program, operate, and manage welding robots.

They will also be needed to solve problems when issues occur. And some welding jobs cannot be done by robots. Human welders will be needed for work that is complicated.

Rob Goldiez was the leader of Hirebotics. This company makes welding robots. He said, "[Robots] take over the dull, dirty, and dangerous tasks, allowing welders to focus on more creative and complex work."[7] By working together, human and robot welders may make the job safer. They might also make welding more efficient.

NEW TECHNOLOGIES

Welders will need to learn how to use new tools and technologies. Artificial intelligence (AI) is one such tool. AI is technology that can do complex tasks. It can write. It can process large amounts of data. It can find patterns. It can even learn and improve over time. Only humans used to be able to do these tasks.

AI can make welding faster and safer. It can also make welding more accurate. AI can analyze designs and materials. It can determine the best way to weld a certain piece. This saves time. It can also help workers make better welds.

AI can watch the welding process in real time using sensors and cameras. It can quickly adjust a tool's temperature and speed. It can spot problems early. These problems include cracks. They also include parts that are not lined up correctly. AI can even predict when welding machines need repairs before they break down.

Welders will also need to learn new techniques. One such new technique is friction stir welding. This technique uses friction to heat and mix metals.

Laser welding allows welders to join small pieces with high precision.

Laser welding is another new technique. It uses a powerful laser beam. The laser melts metal precisely.

Electron beam welding is an older technique. But welders are learning new

Electron beam welding was invented in the 1950s, but new advancements in the method are still being made.

ways to do it. Electron beam welding uses tiny particles to heat metal. It creates very strong welds.

Learning new methods will allow welders to work with a wider range of materials. They will also be able to make better welds. Welders who learn how to use these new techniques will have many opportunities in the future.

Austin Hargett has been a welder for 14 years. He understands the importance of learning new skills. He says:

> *Welding machines are changing. Technology is changing. My job is a constant evolution. As a welder, I believe that I need to learn something new every day.*[8]

GLOSSARY

aerospace

related to the sky and outer space

fabricated

built or created

fumes

gases such as smoke that can be harmful to breathe

industry

a group of businesses that each serve a similar purpose

infrastructure

structures such as roads, bridges, and buildings that help a community work

precise

very exact and accurate

scaffolds

platforms people stand on when they need to reach high places

stamina

the ability to keep doing something for a long time

unions

groups of people, especially those in the same field, with common goals

SOURCE NOTES

INTRODUCTION: WELDING FOR OUTER SPACE

1. Quoted in Lauren Sarmir, "Welding to the Moon and Back," *Northrop Grumman*, n.d. www.northropgrumman.com.

CHAPTER ONE: WHAT DOES A WELDER DO?

2. Jo Mackiewicz, "Only 5.3 Percent of Welders in the US Are Women. After Years as a Writing Professor, I Became One," *Conversation*, November 5, 2024. https://theconversation.com.

CHAPTER TWO: WHAT TRAINING DO WELDERS NEED?

3. Quoted in "How to Get a Welding Job: It Starts with Training," *Northeastern Junior College*, September 28, 2021. www.njc.edu.

4. Quoted in "How to Get a Welding Job."

CHAPTER THREE: WHAT IS LIFE LIKE AS A WELDER?

5. Quoted in "Interview 19: Janet, a MIG Welder," *Women Who Weld*, January 16, 2025. https://womenwhoweld.substack.com.

6. Quoted in "Interview 19."

CHAPTER FOUR: WHAT IS THE FUTURE FOR WELDERS?

7. Quoted in Kyanne Peek, "How Welding Automation Is Changing the Welders Role," *Hirebotics*, June 24, 2024. https://blog.hirebotics.com.

8. Quoted in Eugene Demaitre, "ABAGY Finds That Many Welders Welcome Robotics, Looks to 2023 Manufacturing Trends," *Robotics 24/7*, January 27, 2023. www.robotics247.com.

FOR FURTHER RESEARCH

BOOKS

Cynthia Kennedy Henzel, *Vocational Careers in Construction*. BrightPoint Press, 2026.

Stuart A. Kallen, *Exploring Jobs in the Skilled Trades*. ReferencePoint Press, 2023.

Greg Robison, *Exploring Building Trade Careers*. Lerner Publishing Group, 2026.

INTERNET SOURCES

"Learn About Being a Welder," *Indeed*, December 11, 2025. www.indeed.com.

"The 5 Basic Weld Joint Types Explained," *Universal Technical Institute*, September 2, 2020. www.uti.edu.

"What Does a Welder Do?" *Career Explorer*, n.d. www.careerexplorer.com.

WEBSITES

American Welding Society

www.aws.org

The American Welding Society is a nonprofit organization dedicated to promoting the welding profession and supporting those who work in the industry. Its website contains resources to help welders get certifications, education, and scholarships.

SkillsUSA

www.skillsusa.org

SkillsUSA is a national nonprofit organization for students preparing for careers in technical, skilled, and service occupations. The organization provides students with scholarships and resources to further their careers.

Women Who Weld

www.womenwhoweld.org

Women Who Weld was created in 2014. It is a nonprofit organization that teaches women how to weld and supports women in the welding industry.

INDEX

IMAGE CREDITS

Cover: © Merch Hub/Shutterstock Images
5: © Halfpoint/Shutterstock Images
7: © Futura Images/Shutterstock Images
9: NASA Images
11: © Thang Nguyen Photo/Shutterstock Images
13: © Thaweesak Thipphamon/Shutterstock Images
14: © Ushi Torfs/Shutterstock Images
17: © Evgeny Tkachev/iStockphoto
18: © kadirkaba/iStockphoto
21: © ViewApart/iStockphoto
25: © vm/iStockphoto
26: © Crs Photo/Shutterstock Images
29: © Phynart Studio/iStockphoto
30: © Phynart Studio/iStockphoto
33: © Phynart Studio/iStockphoto
37: © Dny59/iStockphoto
39: © wi6995/Shutterstock Images
40: © Everyonephoto Studio/Shutterstock Images
44: © Salman Timur/Shutterstock Images
47: © 24K-Production/Shutterstock Images
48: © N_Sakarin/Shutterstock Images
51: © Suney Munintrangkul/Shutterstock Images
52: © Amorn Suriyan/Shutterstock Images
55: © genkur/Shutterstock Images
56: US Department of Energy

ABOUT THE AUTHOR

Carla Mooney is a graduate of the University of Pennsylvania. She is the author of many books for young people.